—THE·
LEEDS PALS

STEPHEN WOOD

AMBERLEY

First published 2014

Amberley Publishing
The Hill, Stroud
Gloucestershire, GL5 4EP

www.amberley-books.com

British Library Cataloguing in Publication Data.
A catalogue record for this book is available from the British Library.

ISBN 978 1 4456 1945 3
E-book ISBN 978 1 4456 1963 7

Typeset in 10pt on 12pt Sabon.
Typesetting and Origination by Amberley Publishing.
Printed in the UK.

Introduction

The year 1914 rings through many hearts as one of the most infamous years in British history. It was the outbreak of the greatest war in living memory, the First World War. The origins of this war lay deep amid early twentieth century European politics; however, it is the history of the men who were tragically affected and the stories these men have become famous for which will be told in this book.

Britain in the early stages of the twentieth century had nothing more than an army that was used to police its vast empire; the regular army had become insufficient to fight a war that had been brewing for numerous years prior to 1914. It was therefore the brainwave of Field Marshall Lord Kitchener, who was at the time the Secretary of State for War, that the Regular Army Britain maintained would not be substantial enough to wage a war this size.

Kitchener's first day in office saw much change in the attitude of many within government towards the contemporary army and of course their vast navy: there was the approval of 500,000 men to join the regular army and of course the famous 'call to arms' the following day, followed by a speech in Parliament which ultimately outlined his impression that the current army could not have sustained a war in Europe. It was then decided that each regular regiment were to have within them a 'service battalion'. These were to be made up of enlisting volunteers, as opposed to the standard method of compulsory service; this gave further credence to Kitchener's argument, especially when thousands answered his call to arms in the days after this announcement.

This period saw the British government unable to cope with a surge in public demand to join the war effort. Kitchener relied extensively upon local parishes and councils to organise the enlistment of men for their particular territory. This was all leading towards the formation of what were to become famous as the Pals Battalions.

Volunteers take the oath of allegiance at Leeds Town Hall in September 1914, their first step in becoming a Leeds Pal.

Opposite: Field Marshall Horatio Herbert Kitchener. He became Secretary of State for War and one of the most recognisable men of the twentieth century, featuring as the figurehead for recruitment in the First World War.

It was essentially Lord Derby of Liverpool who brought it to the attention of Kitchener that men were more likely to enlist if they could join up with their friends, family, work colleagues and neighbours. From this came the first Pals Battalion, the Liverpool Pals. Derby's success soon hit the *Liverpool Daily Post* and subsequently news travelled throughout the north of England. Liverpool was followed by Manchester, Birmingham and other notable cities in England; soon it would be the turn of Leeds to raise its own Pals Battalion.

Right: A poster that ran alongside Kitchener's Call to Arms. It was produced to tell would-be volunteers the kind of men they were looking for and the commitment they would have to give; many would volunteer forthwith.

Below: Leeds Pals cap badge.

Opposite: Kitchener's famous 'Call to Arms' poster which was spread throughout the country in 1914 to drive recruitment and enthusiasm for the war.

Your King and Your Country Need You.

A Call to Arms.

An addition of 100,000 men to His Majesty's Regular Army is immediately necessary in the present grave national emergency.

Lord Kitchener is confident that this appeal will be at once responded to by all those who have the safety of the Empire at heart.

TERMS OF SERVICE.

General Service for a period of 3 years or until the war is concluded.

Age of enlistment between 19 and 30.

HOW TO JOIN.

Full information can be obtained at any Post Office in the Kingdom, or at any Military Depot.

God Save the King.

Recruitment in Leeds

Adorning the streets of Leeds on 3 September 1914, the banners read 'BUSINESS MEN show your patriotism your country needs you'. At 9 p.m. hundreds of well spirited men of all ages and sizes crowded the steps of Leeds City Hall, waiting to enlist in the Leeds Pals Battalion. It was soon noticeable that the criteria to join the Leeds Pals were very different to that of other battalions.

It was not only the banners and posters around Leeds that were used to call up willing participants; in coordination with the Tramway Depot they were able to run a recruitment tram, decorated with all manner of posters and lights, which wound its way through the streets of Leeds recruiting all whose eyes glared upon it. Adding to this recruitment drive, the local press printed the lines 'only non manual workers between the ages of 19 and 35 should apply'. Worthy applicants were turned away; one such young man, E. Robinson, recalled that on arriving at the town hall he was asked the occupation of his father, to which he replied, what did it matter what his father did, he wasn't joining up. After some time he remarked that his father was a farm worker; unfortunately, he was told that only professional men's sons or those whose fathers had businesses could join. The exclusivity attached to the battalion drove enthusiastic applicants away to other battalions but it was this class distinction that set this battalion apart. The criteria of intelligence and education were to denote the membership of this 'feather bed battalion'.

This phrase was coined by one writer in a bitter letter which explained how the lord mayor should not forget to equip the battalion with dressing gowns, slippers, eiderdowns, whiskies and a few billiard tables.

The fundamental notion of the Pals Battalion was to recruit friends and colleagues from the same middle class social circles. The following day half the required men had volunteered, and on 10 and 11 September the enlisted Pals were to attend the rigorous medical examinations.

The desperation of some men to go to war with their friends grew too much and one Pal, Morrison Flemming, on entering the examination room appeared a little taller than usual, after filling his shoes with paper and fastening makeshift heels to them. This was soon apparent to the examining doctor, but after pleading with him the doctor allowed him to join up. The early twentieth century saw the outbreak of tuberculosis and there sprang up tuberculosis hospitals, but it was to the shock of these hospitals and the people who knew of this that at the outbreak of war many of the beds in these buildings were empty. Many of the patients slipped into regiments throughout England, including the Leeds Pals, but many sadly died of their disease. It is stories such as these which enable us to understand the attitudes that the young men had to war.

The Leeds Pals raising committee stand proudly outside Leeds Town Hall. The cricketers Dolphin, Kilner and Lintott were among them to bring something of a celebrity status to the Leeds Pals and would help to gain much popularity for the battalion.

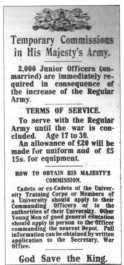

Above left: A Leeds Pals recruiting poster used in September 1914 calling for all businessmen to come forward; these we know were the kind of men that were to be attracted to the Pals.

Above right: This poster is similar to the other Call to Arms poster, but this one is asking for suitable men to hold temporary commissions in the battalion and sets out the criteria for such roles.

'The Featherbed Battalion' men carry their mattresses and bedding to their tents on arriving at camp in Colsterdale.

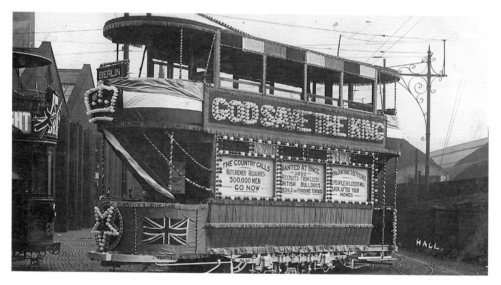

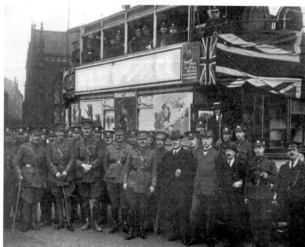

Above: The famous Leeds Pals recruiting tram that toured the streets of Leeds, dressed in all manner of patriotic symbols.

Lt-Col. S. C. Taylor, Major L. M. Howard and Dr White posing outside the recruiting tram that would make its way through the streets of Leeds; it helped to raise over 800 volunteers on 15 June 1915.

Morris Bickersteth can be seen as a fine example of the patriotism of the time. Bickersteth was able to memorise the letters on an eye testing card, as he knew that being partially sighted in his left eye would severely affect him joining the battalion.

It was at a meeting on 4 August 1914 with the intention of organising the panic that had struck Leeds in the days following the declaration of war that Lord Mayor Edward Brotherton proceeded to offer to put half his capital at the disposal of his country. With these funds each man would be given a cap badge featuring the Leeds coats of arms (officers were to receive a silver cap badge) as well as military hardware from the War Office. His gesture to the city of Leeds was estimated to be around the sum of £6,000, which in 1914 was a substantial amount of capital.

Above: Some of the Pals being measured for their uniforms, which were paid for by the Lord Mayor, Edward Brotherton.

Right: Lord Mayor Edward Brotherton generously gave much of his personal fortune to the war effort.

Sportsmen

It was not only middle class men such as business owners' sons and lawyers that were to enlist into the Leeds Pals; Great Britain provided its finest, and for Leeds this meant notable sportsmen of the time. Among these famous idols were Evelyn Lintott and Morris Flemming, both footballers of their day, as well as prominent cricket players Roy Kilner, Major Booth and Arthur Dolphin, but it was not just team sports players that offered the only possibilities; athletes such as George Colcroft and Albert Gutteridge were also willing to be part of something so unique it may never be seen again.

Corporal Roy Kilner was born in Wombwell in South Yorkshire and was aptly described as Friar Tuck of the Yorkshire cricket team. With this in mind, he was without doubt the most popular player to wear the great white rose cap.

For Kilner, cricket ran through his blood, along with his passion to fight for his country come 1914. His uncle, Irving Washington, was a stylish left-hand batsman who played 44 matches for Yorkshire; following in his uncle's steps, Kilner started out his career playing for Mitchell Main, a local side, and by 1910 had made the Yorkshire second team. Kilner played four seasons for Yorkshire, mainly as an aggressive left-handed batsman whose preferred strokes were the off-drive and pull. His greatest season as a batsman was that of 1913, when he scored an unbelievable 1,586 runs at a 34.47 run rate, one of ten occasions when he exceeded 1,000 runs during a summer. This magnificent player was definitely a credit to the Leeds Pals battalion and joined alongside hundreds of other young chaps from the Leeds area, all of whom knew who this great cricket idol was, but it was not until after the war that he exceeded this status.

Kilner was one of the very few lucky men who lived through the war and it was the death of his good friend Major Booth that brought new duties for him in the 1920s. Four times he completed the 'double', with a best of 1,404 runs and 158 wickets in 1923. His nagging bowling, with its variations of pace and flight, made him a vital component of Yorkshire's championship winning side of the early 1920s and in 1924 he was selected as one of Wisden's Five Bowlers of the Year. Kilner toured Australia in 1924/25, assisting in England's Test victory at Melbourne, scoring 74 and taking five wickets for 70 runs in the match. He also toured the West Indies in 1925/26 and in total played nine times for England.

The 1911 Yorkshire cricket team. Kilner is in the back row, third from the left; Dolphin and Booth are seated at the front. All three men joined the call to arms together as friends who had already bonded during their years of cricket together.

Pals in both senses of the word:
cricketers Dolphin, Booth and Kilner.

Kilner pictured in the outfit that made him famous in the batting nets of Yorkshire.

The Bystander, January 7, 1925 30

Our Ash Hunters Down Under

THE M.C.C. TOURISTS AS SEEN BY AN AUSTRALIAN ARTIST

An Australian artist drew these caricatures of the England Ashes team for the *Bystander* in 1925. Roy Kilner, surviving the war, went onto play in this team.

Above all else, though, Kilner is best remembered for his enduring sense of fun, even during the stern conflict of a Roses encounter: 'He says good morning and after that the only thing he says is – How's That?' He became one of the most easily recognised county cricketers of the 1920s with his broad, round face and solid round appearance. His easy disposition won him friends wherever he went. Enteric fever claimed him at the age of only 37, contracted during a coaching trip to India in 1927/28. It is estimated that around 100,000 people crowded the streets of Wombwell for Kilner's funeral.

Later Wisden wrote:

> Few modern professionals commanded such a measure of esteem and kindly regard from his own immediate colleagues and opponents in the cricket field as did Roy Kilner.

All information for Roy Kilner is taken from *100 Greats Yorkshire County Cricket Club*, M. Pope and P. Dyson.

This is a fitting tribute to a true gentleman who not only represented his country on the cricket pitch but in the fields and trenches of war-torn France; not many men can say they have had an experience such as this. He was a true great of his time and one who, along with his friends and colleagues both in battle and on the pitch, will be remembered. He was not alone in his fascination for and love of the sport; there were also notable players such as Arthur Dolphin and Major Booth who also played the sport for their county.

When we look at famous sports idols in the early twentieth century in connection with the Leeds Pals, there is one more man whose story stands out: Evelyn Lintott led his men gallantly into battle with his words 'Now Leeds' echoing through the trenches. This is his story.

Evelyn Henry Lintott was born on 2 November 1883 and became an extraordinary football player. He was a halfback and used his skills with such teams as Plymouth Argyle, Queens Park Ranger, Bradford City and eventually Leeds City, where he would end his career and join the Leeds Pals. He was not just a club player but also represented England in the national team, being capped several times.

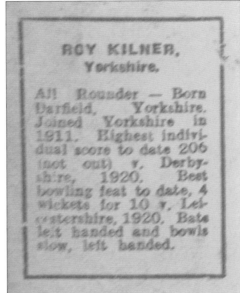

Above left: The front of a cigarette card bearing Kilner's photograph; many young lads would have been collecting these.

Above right: The back of the cigarette card, showing some of Kilner's excellent credentials in the sport of cricket.

Right: The infamous Eveleyn Lintott in his trademark England gear, the shirt bearing the 'Three Lions'. This was one of the many teams he played for in his career.

His career started with a chance to play for Woking and here assisted the club in winning the East and West Surrey League and Surrey Charity Shield before moving to Plymouth Argyle in 1906. However, he only made two appearances for the team before taking the step to move to London to become a teacher at Oldfield Road School in Willesden. Here he picked up his love for the sport once again, and took part in over thirty games for Queens Park Rangers. This clearly sounded him out as a successful player and from here he moved north to the city of Bradford in 1908; after a short career here, he then moved down the road to Leeds, which we now know would be his last club and the last time he would play his favourite sport on the big stage. His career was not only dominated by his love for the sport but also his intellect; he attended the Royal Grammar School in Guildford, which was an independent school for boys, and was able to gain a place at St Luke's College in Exeter, where he studied teaching. Football players during these early years were not full time and many of them had a career as well as their football career. Lintott became a schoolmaster and was also the only professional football player to be commissioned. He was clearly a well-defined and ultimately well rounded individual from what one can only imagine was a privileged background, but like many others he did not let this get in the way and answered Kitchener's call to arms. He gave his life leading his men into courageously into battle in the same manner as perhaps he may have done on the football pitch. On 11 July 1916, the *Yorkshire Evening Post* printed this article.

Lieutenant Lintott's end was particularly gallant. He led his men with great dash , and when hit the first time he declined to take the count. Instead he drew his revolver and called for further effort. Again he was hit and struggled on, but a third shot finally bowled him over. [...] Lintott [was a] gallant sportsman who knew how to die – but then so did all the boys. They went out to almost certain death with the cry 'Now Leeds' on their lips.

Far left: Lintott also had his own cigarette card; in this drawing he is featured in his Bradford City kit.

Left: In this cigarette card, Lintott is featured in another one of his kits from when he played for Queens Park Rangers.

The other uniform that Lintott was to proudly wear come the year 1914; here he is dressed smartly in army uniform, proudly bearing the Leeds Pals cap badge.

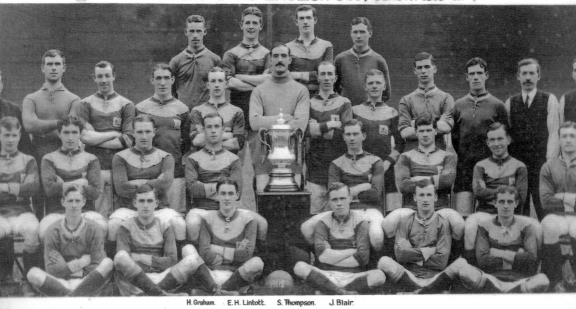

H.Graham.   E.H.Lintott.   S.Thompson.   J.Blair.

urke M.Spendiff. A.Devine. F.O'Rourke. R.Torrance. M.Mellors. J.McDonald. H.Peart. P.Cassidy. H.Hampton. D.Menzies, C.Ha
(MANAGER)   (ASSISTANT TRAINER.) (TRA

O.Fox.   W.Gildea.   R.Campbell.   J.H.Speirs.   G.Robinson.   D.Taylor.   F.Thompson.   F.Farren.

W.B.Walker. J.McIlvenny. P.Logan.   R.Bond.   J.Young.   G.Gane.

A team picture from when Bradford City won the English Cup in the 1910/11 season, featuring the great Eveleyn Lintott with his friends and team mates.

Colsterdale

The time spent at Colsterdale was for most of the Pals the best time of their lives. Colsterdale sits in the heart of North Yorkshire and at the time was used by the Water Works Committee. This was the first place that the Leeds Pals would establish as home. What a home this was; it was serviced by a reservoir and a light railway that was used to bring supplies and, more importantly, the men themselves. In 1914 part of this area was used by the men that were working on the nearby reservoir but once the resolution had been passed by Alderman Willey and the council to use this area for the war effort these workmen soon left, leaving behind their navvies' huts, which were soon taken over by many of the Pals. The rest would have to survive under tents until they could build huts and substantial accommodation.

The Advance Party, as it was to become known, comprised 105 men, one of them, now a sergeant, Evelyn Lintott the professional football player. They set off from Leeds at 7:51 on the morning of 23 September 1914 and they were to be accompanied by the stores and other essential necessities to Colsterdale, to ready the camp in preparation for the rest of the men. Much of their time was spent erecting the white bell tents that were to be used as temporary accommodation for the men of A, B and C companies; D company were lucky enough to be housed in the workmen's housing.

On the 23rd, the Leeds Pals were ordered to attend a gathering at the town hall; while there they were given their kitbag with essential supplies, and also a card which detailed which company they were to join. Two days later, on 25 September, they set off from Leeds by train with a crowd of over 20,000 people waving them off.

After the journey to Colsterdale, some of them having walked some of the distance, they were addressed by Colonel Stead, who was at the time their commanding officer. His speech recognised the high standards which the West Yorkshire Regiment already upheld and the punctuality that these men were to attain. It seemed that punctuality was of foremost importance, and it was hoped that this class of men would know a little something of punctuality in the first place. This, for most, was not an issue and on the first morning all but a few dragged themselves out of their bell tents at 5:45 a.m. to get ready for morning coffee and the hearty breakfast that Colonel Stead had promised them the day before.

The men at Colsterdale working out; they look to be having great fun and making bonds of friendship that would last a lifetime.

A group of Pals receiving parcels from their family and loved ones in 1914.

A group of Pals sat around during down time at Colsterdale, chatting possibly about the day's events. In the background the white bell tents can be seen.

The new huts that were built in Colsterdale; the men were living in bell tents before these permanent structures were built.

An officer clears turf to help pitch in while being watched by his men at Colsterdale.

The guard taking their orders in Colsterdale, pictured with their rifles in hand.

A wet and miserable day in the camp; the men would have been glad to have the huts to sleep in at this point.

Above: The Pals posing with their newly received kit bags at Costerdale.

The crowds of men arriving at Colsterdale in their droves, their tents already set up for them in the background.

Opposite: The first page of orders that the men were given to abide by on arriving at Colsterdale; it was of the utmost importance that this document was abided by, issued by Colonel Stead prior to the men arriving at the camp.

The Leeds Battalion.
The Prince of Wales's Own West Yorkshire Regiment.

ORDERS FOR CAMP
BY
Lieut.-Colonel J. WALTER STEAD, V.D.

TOWN HALL, LEEDS, *September 21st*, 1914.

1. CAMP.

The Battalion will encamp at Colsterdale on **Friday, September 25th, 1914.**

2. ADVANCE PARTY.

Men who have given in their names will attend at Headquarters, on **Tuesday, September 22nd,** between 6 p.m. and 8 p.m., to draw kitbags. It is recommended that those who can do so should bring in a pair of Blankets in a paper parcel for their own use. These will have a label attached with name and number and will be issued on arrival at Camp. Each man will receive a card showing the Company to which he has been posted, and on presenting this to the Quartermaster-Sergeant will receive his kitbag, to which the card will be affixed.

The advance party, under Sergeant-Major Yates, will parade with overcoats, sticks and kitbags, at the North Eastern Railway Station, Leeds, on **Wednesday, September 23rd**, 7.30 a.m., and proceed by train to Masham for Colsterdale Camp.

3. ISSUE OF KITBAGS.

The remainder of the Battalion will attend at Headquarters, on **Wednesday, September 23rd,** between 4 p.m. and 8 p.m., to draw kitbags. Owing to the scarcity of Camp equipment, it is recommended that those who can do so should then bring a pair of Blankets in paper parcel for own use. These will have a label attached with name and regimental number, and will be issued on arrival at Camp. Each man will receive a card showing the Company to which he has been posted, and will receive from the Quartermaster-Sergeant his kitbag with the card affixed.

4. ARTICLES TO BE BROUGHT.

Men will attend in plain clothes, with caps, sticks and overcoats, and each man is recommended to be in possession of the following articles in his kitbag:—Two pairs of socks, one shirt, pair of pants, hair brush and comb, tooth brush, clothes brush, small dubbin brush, razor and shaving brush, two towels, pair of strong leather laces, pair of shoes (canvas preferred) to wear in camp after drill. No unauthorized bags or boxes will be taken to Camp. Hair should be cut short.

5. OFFICERS' BAGGAGE.

Officers should arrange to have their baggage ready packed and labelled at Headquarters, not later than 8 p.m., on **September 24th.**

6. BATTALION PARADE, SEPTEMBER 25th.

The Battalion will parade in plain clothes, with overcoats, sticks and kitbags, at the North Eastern Railway Station, Leeds, on **Friday, September 25th.** Men will fall in by Companies on the platform at 8.30 a.m., and afterwards proceed by rail to Masham for Colsterdale. On arrival at Masham kitbags will be collected and transported to the Camp.

7. ENTRAINING.

When entraining silence must be maintained as far as possible by all ranks to ensure expedition in the work. Men must not entrain until ordered.

8. HOURS OF REVEILLE, &c.

The Buglers will sound the following calls at the hours stated :—

Reveille	5.45 a.m.
Dress for parade	6.15 a.m.
Parade	6 30 a.m.
Retreat	7.0 p.m.
Tattoo, First Post	9.0 p.m.
Tattoo, Last Post	9.30 p.m.
Lights Out	9.45 p.m.

Absolute silence is to be maintained in Camp between the hours of Lights Out and Reveille.

9. SATURDAY'S PARADE.

The Battalion will parade as strong as possible on **September 26th**, at 6.30 a.m., for check roll-call. All N.C.O.'s and Men, no matter how employed, must attend this parade. Employed men will be allowed to go back to their work when permission has been obtained. Pay Sergeants will attend with their pay-lists to check the names.

Colonel Walter Stead, who was
the commanding officer of the
Leeds Pals from August 1914 to
May 1915; he was superseded by
Stuart Taylor.

Colonel Walter Stead addresses the battalion in Colsterdale, explaining that punctuality was
paramount to their role in the army.

The cook houses that were used to feed the 1,100 men while training at Colsterdale. They look makeshift but the cooks were able to create large staple meals for the men.

A field full of the large white bell tents used to house the men until the huts were built.

Their first introduction to army life was a lesson in Swedish drill and it was to be this that would be the outline for choosing NCOs or Non-Commissioned Officers. It seemed that amassing this amount of men had been relatively effortless throughout the country as patriotism and English spirit had gathered such momentum; however these men on the whole had never had any military experience and it was to be this that was to become the challenge. Although they may have not had any military experience, nevertheless the men were sorted into prospective skills, trades and in an attempt to become self-sufficient. Private Clifford Hollingworth recalled in an interview with Laurie Milner in 1988 his experience of this self-sufficient battalion.

> They started picking specialists out. We had a man called Summersgill, he was a plumber, well he was put on to the water. We had a man who was a butcher, he was put into the butchers department. A man who was a tailor was put into the tailoring department to patch up the clothes, and men who had served behind counters were put into quartermaster stores. By the time we got organised after six months, you see, they'd moulded us into a unit, and from the colonel right down to the junior private you've got to have some kind of building.

For men that had been school teachers, Swedish drill was something which many of them had learnt in their training and thus this small amount of knowledge would automatically make them non-paid lance corporals; this then became the beginnings of the regiment's structure.

Their time spent in Colsterdale was made up of differing types of training that of course they would need once they joined active service in Egypt and France. As we have seen, the battalion relied upon the skills of its men for the smooth running of the camp.

Being a cook was a full time position at Colsterdale, having to feed around 1,100 men, making sure they had sufficient food to keep them active and healthy. In the *Yorkshire Evening Post* on 4 December 1914 appeared a small article outlining the provisions the men in the cookhouse would have to make available in a two-day period, still of course being provided by the concessions of the Lord Major at the time, Edward Brotherton.

Sunday
6.00am, 150 gallons, coffee;
7.45am, 500 gallons coffee, 500lb Sausage;
12.30pm, 1,000lb Beef, 400lb Puddings, 9 sacks Potatoes;
4.30pm, 200 gallons tea;

Monday
6.00am, 150 gallons Cocoa,
7.45am, 200 gallons Cocoa, 500lb Haddocks;
12.30pm 1,000lb Beef, 120lb Peas, 9 sacks of Potatoes;
4.30pm, 200 gallons of Tea.

Five men from a trench digging party with their spades, looking rather impressed with their pipes in their mouth.

The Pals on a regimental route march to Middleham with their regimental pet, Belgie.

The Pals lay down in their 'trench', using the railway banking to practice their attack.

A group of friends in civilian dress posing for a picture, some with a drink in their hand.

The men gallantly attacking Hill 60 in Colsterdale, preparing for the day that would become 1 July 1916.

Pals in their uniform taking part in their daily training.

The men preparing for an 'attack' in the fields of Colsterdale.

This is just a small insight into the varied meals that were available. This was a time of war and civilians in the towns and cities, many struggling to pay their way having lost members of their family to the war effort, would have perhaps dreamed of such meals. Colonel Stead on their first day did mention there would be coffee and boy was he right.

Once at Colsterdale it was not solely about training and drill and taking part in chores around the camp; down time played a large part in the success of this battalion, becoming not only colleagues but a large close knit family with similar values and expectations. For the sportsmen in the camp, some of whom have already been mentioned, a sports and recreation committee was set up. They took part in all different kinds of sport, boxing, cross country and football to name a few, and even held a sports day on 12 May 1915 attended by the Lady Mayoress, Una Ratcliffe, who gave out various prizes and awards.

It is apparent that manual skills and sporting skills were prevalent within the Leeds Pals and both were used in their own way to achieve certain goals. However, what if you had neither but instead a passion for the arts, drama or even magic? You'd need not worry.

'The Owls' as they were to become known was the place for you. They were the amateur dramatic society or committee; call it what you will, but it was their job to put on shows and pantomimes to ensure the spirit of the men was always at its highest. One of their shows took place in France, on 28 December 1917, when the lads had returned from the trenches to have their own Christmas. It must have felt as though they were in another world; there was a huge marquee set up, Christmas dinner with all the trimmings and of course *Aladdin*, the pantomime that the Owls had chosen to perform on this special day.

Once their time in Colsterdale had finished, they spent a few days in Fovant in Wiltshire, in a large purpose-built training ground, before heading to Egypt. The idea of sending the men to Fovant was to enable them to practice large scale manoeuvres in terrain similar to that of France. These large scale attacks were organized with other regiments and so gave a feel of the large scale battles they were to one day take part in for King and Country. This kind of training could not be done in the uneven, hilly terrain of North Yorkshire.

One of the competitions featured in the Colsterdale sports day; this event sees two men taking part in blindfold boxing, being watched by a large crowd.

The teams that took part in the inter-platoon rugby competition, the winners wearing horizontal hoops.

The sprinting race at the sports day event.

Sports players at Colsterdale at the prize-giving ceremony after the sports day they held on 12 May 1915. The day included many sports across different fields.

The winners being presented with their prizes by the Lady Mayoress of Leeds, Una Ratcliffe.

'The Owls' in their other outfits beside their army uniform before performing to the men. The Leeds Pals emblem is proudly displayed in the background.

'The Performance' - the men would have certainly enjoyed this.

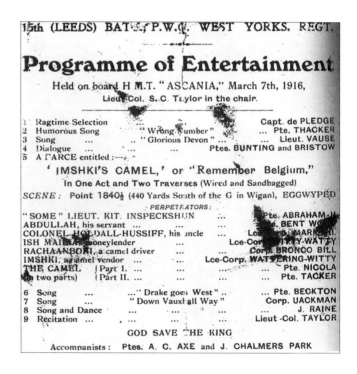

The programme for the show on board HMT *Ascania* performed by 'The Owls'.

HMT *Ascania* was a Cunard steamer, which ordinarily sailed to Canada. During the war she operated as a troop ship and as an armed merchant cruiser, keeping the sea lanes open. She was sunk in 1918 off Cape Ray, Newfoundland.

Troops boarding the HMT *Ascania*.

Life aboard the HMT *Ascania* was similar to ashore, with physical drill and other training to keep the men fit and ready on the voyage.

Kit inspection aboard the HMT *Ascania* on one of her trooping voyages.

Soldiers' kit bags being stacked after making the journey from Colsterdale in North Yorkshire to Fovant, Wiltshire, where the Pals were to be stationed for further training until their departure overseas.

The picture looks towards to the main entrance of the camp at Fovant. The picture shows the sheer size of the training ground; it is understandable why this place was chosen to undertake these large scale manoeuvres.

Quartermaster R. J. Anderson, Captain P. H. Mellor and Major L. P. Baker doing the rounds during an inspection of the camp at Fovant.

Lt-Col. S. C. Taylor with the regimental band, posing for photographs in Fovant. They put on shows for the men and played when the Pals marched.

A smartly dressed squad on parade on one of the large training areas in Fovant.

Two Pals during Lewis gun practice in Fovant.

IN HAVERSACK

EQUIPMENT

IN FULL FIGHTING ORDER AND EQUIPPED FOR EVERY EMERGENCY.—THE BRITIS

THIS GRAPHIC PORTRAYAL OF EVERYTHING AN INFANTRYMAN CARRIES ON ACTIVE SERVICE WILL ASTONISH MANY WHO HAVE ON
JACKET SERVED OUT TO OUR TROOPS IS AN ADDITIONAL ITEM, AN

IN KNAPSACK

ON PERSON

SOLDIER'S BURDEN IN THE FIRING-LINE PICTURED FROM A TO Z.

VAGUE IDEA OF THE QUANTITY OF ARTICLES INCLUDED IN THE EQUIPMENT OF A SOLDIER OF THE LINE. THE WINTER GOATSKIN
OMETIMES HE CARRIES ALSO EXTRA RATIONS AND FUEL.

Mortar practice; they played a key part during the barrage prior to 1 July 1916.

Trench digging in Fovant, to be used to practice assaults.

Sat in the makeshift trenches which were dug in Fovant, waiting to go over the top. This took place not only for physical practice but to understand the mental process behind going over the top into No-Man's Land.

Men in the trenches
practicing trench warfare.

Practising the assault over
the top into No-Man's Land
in Fovant.

A fine example of how the
battalion were to provide
for themselves. Private
Temperton was in charge of
the battalion pig farm.

Egypt

The British realised that the decision in December 1915 to evacuate the Gallipoli Peninsula would release Turkish troops for operations elsewhere. The Turks had already attempted to take control of the Suez Canal in February 1915, hoping to gain support from Egyptian nationalists opposed to British rule.

With this in mind, it was decided to stiffen the defences around the canal. Rumours that the Pals were going to France were soon dispelled when the transport section of 102 men, under the command of Captain Boardall and Lieutenants Smith and Everitt, left Devonport on 6 December on board HMS *Shropshire*, bound for Egypt.

The remainder of the battalion, plus their pioneers (12th KOYLI) and Army Service Corps personnel, left Liverpool along with the rest of 93 Brigade (6,000 troops) on board the *Empress of Britain* (a converted liner) on 7 December for the same destination.

The *Shropshire* arrived at Port Said on 20 December after a fairly rough but uneventful crossing. The rest of the brigade on board the *Empress of Britain* had a more troublesome voyage. While zig-zagging in an attempt to discourage enemy submarine attacks, she collided with a French mail ship, the *Dajurjura*, and had to put in to Valletta harbour to enable repairs to be carried out. This is an extract taken from a letter by Private Robert Norman Bell cited in Laurie Milner's *Leeds Pals*.

> For a moment the vessel rocked from side to side causing hundreds of enamel plates piled on the mess tables to slide to the deck adding to the alarming effect of the collision. The silence that followed was only broken by the shuffling sound of hundreds of men sliding out of their hammocks and feeling around for their life jackets in the almost complete darkness... after a short while our Company Commander came down the companion and in his well know confident tones addressed us: 'It's alright men, we have only run into a fishing smack.'

Leaving Malta three days later, she arrived safely at Port Said on Tuesday the 21st after a brush with a submarine that fired at least one torpedo in her direction.

The Pals disembarked on the 22nd and marched to No. 8 camp, where they stayed acclimatising until the 30th, then moved 32 miles up the canal to guard various points in the desert. The men were relatively unscathed after their time in Egypt; on the whole it was small skirmishes and mainly active guard duty in the areas to which they were posted. Some of the men in local villages and towns did not take kindly to the Pals visiting but it was Christmas 1915 when Clifford Hollingworth and a few friends defied orders to go into town and took the trip in their stride.

Private F. Robinson outside his dugout in in the desert, dug into the sand and surrounded with sandbags to protect them from desert conditions.

A beautiful view of the Suez Canal, which the Pals were sent to help defend in 1915. In the distance is El Kantara, where they were stationed for a period of time.

The view of the battalion camp, showing the distinctive bell-shaped tents in the distance.

Hellewell, Smith, Tomlin and Child filling sandbags in the desert to use as defences and to build dugouts with. Clearly, getting hold of sandbags was not an issue in this environment.

Some of the men doing their everyday cleaning within the camp.

Captain Mellor with some of A Company, pictured outside the tents in one of the camps in the desert.

Camels weren't the only method of transporting goods; in this picture we see a battalion transport wagon loaded with goods from the base.

Turkington and Macauley pictured with camel transport; the camel is carrying large water tanks, which were a necessity in the hot desert conditions.

Captain Whitaker overseeing the unloading of the regimental stores.

A group of camels used to transport goods and supplies between camps in the desert with an armed guard. They would have been an easy target for attack by Turkish troops.

Captain Mellor interrogating an Arab in the desert, perhaps for information relating to enemy movements.

The fatigue party resting in the desert sun. Fatigues are work such as cooking, cleaning and other chores; men were known to be put on fatigues as a form of punishment.

A military cemetery in the desert; the grave is that of Private Wintle, sadly the first man in the Leeds Pals to give his life.

This camp in El Dueidar where the men were stationed was close to an area where an entire Turkish force was defeated in 1916.

Private Catterick in charge of the officers' make-shift cooking range in the middle of the Egyptian desert.

THE EMPRESS OF BRITAIN (C.P.R.) LIVERPOOL

Above: Canadian Pacific Line's RMS *Empress of Britain* was the sister of the ill-fated *Empress of Ireland*, which sank with the loss of over 1,000 lives in the St Lawrence in May 1914. She took many of the men to Egypt in December 1915. She was damaged when she collided with a smaller French ship and had to make for Valletta, Malta, for repairs.

Opposite, above: The *Empress of Britain* leaving Liverpool for Montreal prior to the First World War.

Opposite, below: The sumptuous First Class Library aboard the *Empress of Britain*.

Gibson, Boardall, Willey, Vause and Rayner relaxing in the officers' smoke room. The officers had their own areas where they could relax and which were slightly more comfortable than the other ranks' mess.

The regimental canteen where the men would eat and perhaps share the tales of their day.

While stationed in Egypt, the men lived in make-shift camps. This is a shot of the field kitchens that the cooks used to supply the men with their daily rations.

We went in just for excitement and we saw some dancing girls and we had a drink of wine and we ate one or two nasty dates. Then I forget now who it was, he was a big lad, he said 'I think we'd best be going lads, I think we ought to get out of this place.'

By late February 1916, the threat of invasion along the Suez Canal had decreased to such an extent that on 1 March, after 3 months of sun, sand, and a few minor skirmishes, the battalion returned to Port Said, gathered up their equipment and stores and marched on board the troopship HMT *Ascania* for the journey to France. It was seen on the whole as a luxury voyage considering the hot desert conditions of Egypt. Much of their time on board was taken up by relaxation and makeshift entertainment; this included boxing matches, wrestling events and regimental band performances to pass the time on board the ship. These gathered huge crowds and the atmosphere looks grand. It was sure the last time the men would see these luxurious conditions for some time and many would not see them again. Next stop was the Western Front and the perils of life in the trenches.

Right: Belgie, the regimental pet, relaxing in the hot desert sun.

Below: A camp on the Turkish side of the Suez Canal.

The picquet leaving the camp in Egypt for the night; the term picquet refers to the men who would be stationed in a forward area so as to warn of enemy advances.

The men taking part in one of the many pastimes, wrestling, being watched by many spectators.

The regimental band keeping the men's spirits up during the Mediterranean crossing.

Liversedge, Willey, Vause, Hutton and Foster with Lt-Col S. C. Taylor.

Tom Willey, who was to lose his life in France, is pictured with R. H. Tolson on deck.

A group of men sat on the decks of their transport ship while crossing the Mediterranean.

A group of men posing for a photo on the decks in the Mediterranean.

Wrestling was not the only sport the men took part in during their rest; boxing was also popular and attracted huge crowds.

1916: The Somme

At precisely 7:39 a.m. the men of the Leeds Pals battalion lined up in the trenches with brothers, comrades and friends, many of them holding each other and saying their last goodbyes; the atmosphere must have been immensely strange and unreal and something which cannot be portrayed in any other way. The barrage had stopped and silence reigned over the valley of the Somme and all that would have been heard were the loud whistles of the commanding officers to denote the climbing of the ladders. One Pal, in an interview after the war, recalled that when the eight-day barrage had ended and it was close to going 'over the top', the larks started singing. They strolled into No-Man's Land, the area that divided the German and British lines, zig-zagged with barbed wire. They were met with heavy German machine gun fire, and sadly, for many these would be their last steps. Private Arthur Hollings of D Company recalled that fateful morning and was published in the *Yorkshire Evening Post* on 15 July 1916:

> Mr Willey passed down the order 'Get ready 13' as casually as though an ordinary parade. We then filed out, up the scaling ladder, through the gap in our own wire… No sooner had the first lot got over the parapet than the Germans opened up a terrific bombardment, big shells and shrapnel.

Once all the men had courageously mounted the ladders and entered No-Man's Land, Hollings recalls that the platoon commander, Willey, 'jumped up, and waving his revolver, shouted, "Come on 13. Give them hell."'

Arthur Hollings further recalls his experience of that day.

> No sooner had the first lot got over the parapet than the Germans opened up a terrific bombardment, big shells and shrapnel.

Hollings goes on to write how inspired him and the rest of the lads were when the first lot climbed those ladders and stood in No-Man's Land. It was this sheer determination of the British will to survive that many of these lads had deep within themselves, and no more clearly was it seen than 1 July 1916.

The campaign in the Somme valley, thought up by Anglo-French planners, was designed to draw the Germans south, away from the Verdun area, where the French were being mercilessly pounded on three sides. The war by this time had become a war of attrition; both sides were dug in with no major offensives, except small skirmishes up and down the lines. With this in mind, it was the idea of the German general staff to try and bleed the French army dry. There were huge casualties on both sides at Verdun; nearly one and a half million French and German soldiers died here.

Men on their way to take up positions on the front line.

A working party carrying boards through to the front trenches; they worked at night to avoid detection from German forces.

The view of the barren No Man's Land from the trenches.

Above: British troops posing in the obliterated German trenches in 1917; the sign jovially says 'The Old Hun Line', showing the spirits of the men after taking the German trenches.

The men in the cramped environment of the trenches, in their full uniform.

Above left: A man poses for a picture with his pipe in the muddy trenches.

Above right: A typical trench setting, with ladders, sandbags and a selection of trench mortars laid around.

'No Man's Land'

'No Man's Land'

The devastated 'No Man's Land'

The battle of the Somme was originally planned for August 1916 but due to increased pressure from the French, it was brought forward to 1 July 1916. It was preceded by an eight-day, constant British artillery barrage on the German lines. This was to destroy all signs of life and the immense barbed wire defences in and around the German trenches, allowing the attacking British troops to stroll across No-Man's Land and on to Germany. As we now know, the German trenches were dug deep and were too advanced to be fully destroyed by the British barrage but at the time nobody was to know this and sprits were high, knowing that this eight-day barrage had come to an end and it was now time for zero hour. To lessen the nerves and the worry, many senior ranks provided footballs for the men to kick about as they advanced towards to the German forces.

The Pals were able to put some of the knowledge gained in the large training grounds of Fovant into practice at the Somme, where they were fighting alongside other battalions in their section of the trenches. The Leeds Pals were at the forefront of the charge, leading 93 Brigade into the battle along with another Pals battalion, the Bradford Pals, and the 18th Durham Light Infantry. Their main objective, which was mapped out for the men to use as a reference as well as miniature models of the battle ground, was to join with 94 Brigade. 94 Brigade was made up from three other Pals battalions, Sheffield, Accrington and Barnsley, and would take the village of Serre itself, while 93 Brigade would work their way around Serre and attack Puisieux au Mont. Many of the Pals roughly knew the route and plans of this day, however, as Private Clifford Hollingworth explains, among the chaos of German machine gun fire, mortar fire and barbed wire entanglements this quickly becomes ignored.

We knew about it before, because they took us out into the country behind the lines and they had it all threaded out ... They had red lines, blue lines, tapes and they drew a map by tapes and pegs of our objective. Now when you get to that red line, wait until the next one. When you get to the blue line, wait. This was alright in theory by the headquarters, but when you come to do anything in practice, theory goes out of the window.

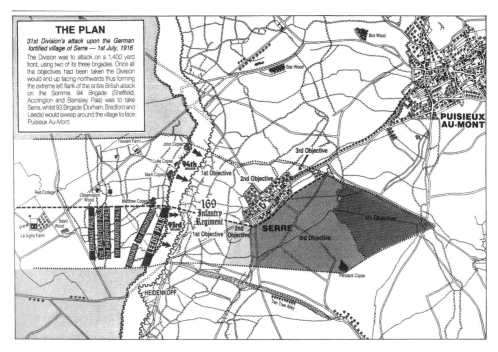

The plans that were made in preparation for the attack on 1 July against Serre. The plans shown were for 31st division, which the Leeds Pals were a part of.

The view the Pals would have from their trenches in April. It gives the feel of how large No Man's Land was and how open this area was.

The weeks prior to the attack, working parties courageously went out into No-Man's Land under the cover of darkness to map out the routes the men would be taking when 'Zero Hour' came on 1 July. One job in particular was to mark out the enemy trenches with coloured ribbon; the reason for this became evident when the men were given matching coloured ribbon so they knew which part of the trench they were to attack. The men would easily become disorientated in the chaos that would ensue on this day, especially considering that in parts the distance between both sets of trenches was one to two miles. Private George W. Cosby recalled his traumatic experience of 1 July in his diary, taken from *Leeds Pals*, by Laurie Milner.

> Wounded about 9am. Bandaged myself up best possible and crawled through trenches in search of a dressing station. Wounded in face en route. Progress very difficult. Intense shelling of support trenches. Found 15th W York dressing dugout and was bound up by Ward the orderly. Afterwards crawled to Basin Wood and had rest.

Tom Willey and his platoon led the attack on the German trenches and his story is particularly sad and very moving; it shows how these men became a more of a family than a unit and the lengths they would go to to help and protect each other. Willey was hit in No-Man's Land and lost both his legs; his body was sadly never recovered. That night some of his platoon went out and searched the area to try and find him, but to no avail. In a letter written after the attack by Arthur Hutton to Willey's parents, he explains just how loved and missed this great man would be and further shows how close many of the Pals became.

> Dear Mr Willey, I cannot possibly express to you my very deep sorrow and regret at the news of Toms death. Everyone loved Tom, as you know, he was more often than not the very life and soul of our mess. He was a great officer and a great man and we shall mss hum tremendously. He was always cheerful and full of life. I can't realise what has happened. All the best have gone. I cannot see why I didn't go with them. I hope both Mrs Willey and yourself will accept my most heartfelt sympathy in your great loss, and find consolation in the thought that Tom, like the boy he was always did his duty. Yours very sincerely, Arthur Hutton.

Medical orderlies rescue the injured from No Man's Land as shells hit close by.

A German trench completely destroyed by British forces. The Pals, after the eight-day barrage, were hoping to stroll across No Man's Land to see all of the trenches in this state. Although this was not the case, the picture does show how some German trenches were destroyed.

Tom Willey (centre front) with No. 13 platoon. Sadly, he did not survive the war.

A roll call on 1 July 1916.

It is understood that during the lead up to the attack on Serre, the German forces had a feeling that a large attack on their lines here would be taking place at some point and therefore were able to concentrate the majority of their efforts on the counter-attack. The attack was made particularly difficult by the positions of the German machine guns, which, when put into action, were able to cut down whole brigades of men, backed up by effective mortar and artillery fire. There has also been considerable debate throughout the years over the actions of General Haig and his aggressive policies that were put into place during the First World War. These meant in particular that the trenches were not dug in positions of advantage, using high ground or other natural landmarks and features for example. Much of this is still debated today and this will possibly carry on for generations to come; all we know is that the cost that day in lives has never been fully accounted. Out of nearly 58,000 casualties, 20,000 men died. The newly trained 'Kitchener's Army' was almost annihilated. These Volunteers or Pals Battalions, two years in the making, almost ceased to exist after this day.

The Leeds Pals suffered in their attack on Serre. Thirteen officers were killed, with two more dying of wounds, and 209 other ranks, with a further twenty-four dying of wounds. The effect on the families and loved ones cannot be imagined. It was reported at the time that there was not a street in Leeds that didn't have at least one house with curtains drawn in mourning. The survivors of this and later battles came home after the war to receive no counselling or compensation. They just got on with their lives.

A post card from Gande reads: 'This photograph was taken when our battalion marched through just after the bosche [Germans] left, ours was the first british regt in the place.' This appears to be Tourcoing, on the Belgian border.

Bernard Gill recovering in a trench hospital with some other injured men before his death.

No. *WYRC/15/1238*   ARMY FORM B. 104—82.
(If replying, please
quote above No.)

N° 2   INFA....   Record Office,

3rd *April* 1918

Sir,

It is my painful duty to inform you that a report has been received from the War Office notifying the death of:—

(No.) *15/1238*   (Rank) *L/Sgt*

(Name) *Bernard Gill*

(Regiment) *W. YORKSHIRE REGT.*

which occurred *43 Cas Clg Stn France*

on the *25 March 1918*

The report is to the effect that he *died of wounds received in action*

By His Majesty's command I am to forward the enclosed message of sympathy from Their Gracious Majesties the King and Queen. I am at the same time to express the regret of the Army Council at the soldier's death in his Country's service.

I am to add that any information that may be received as to the soldier's burial will be communicated to you in due course. A separate leaflet dealing more fully with this subject is enclosed.

I am,

Sir,

Your obedient Servant,

Colonel, O. i/c

Lance Sergeant Bernard Gill died of wounds; this is the notification that the family would receive. They sent their son, father, uncle and even friends to war in 1914 and many in return received this letter and the painful memories of knowing they will never come back.

Extraordinary stories

As we now know, the Leeds Pals weren't just a battalion of men but 1,000 individuals that come with their own stories, social background and personalities. These are just some of the moving and emotional stories of the Pals. It is important to note that these were ordinary men, boys in some cases, who had led at the time normal lives with families, jobs and other everyday commitments and come 7:40 on the morning of 1 July they ceased to exist. This is their story.

Jogendra Sen

The social changes that occurred in the twentieth century changed thinking and mindsets the world over. Many examples of this can be seen with revolutionaries such as Rosa Parkes, the Suffragette movement, Martin Luther King and many others. It was the views of millions of men and women throughout the world that saw these political and social movements spring up. Racial segregation in all walks of life can define much of the social make up of the twentieth century, especially within the army. This can be seen in the Second World War with the Tuskegee Airmen as well as the First World War. It is evident due to the social make up of the British army that yes there were Indian and black regiments. However, for a man of another race to fight alongside white men in the same battalion was very rare; this is the story of one such man, named Jogendra Sen, who defied the social mindset and bravely stepped forward and joined the Leeds Pals.

Jogendra Sen was the only Indian man to serve with the West Yorkshire Regiment; this alone shows the attitude towards these men in 1914. He was born in 1887, in a place called Chandenagore in India, a single man who took the huge step of travelling to England to attend university in Leeds in 1910. While studying there he attained a Bachelor of Arts degree in engineering despite the fact that he was also working as an assistant engineer at Leeds Corporation Electric Lighting station. He, along with thousands of other men throughout the country, saw the threat to freedom and what was now his adopted home and became the first man to volunteer for the

Jogendra Sen.

Leeds Pals battalion. He was not put off by the racial segregation and the possibilities of neglect but took this in his stride.

The following extract is from an interview with Laurie Milner in 1988.

We had a Hindu in our hut, called Jon Sen. He was the best educated man in the battalion and he spoke about seven languages but he was never allowed to be even a lance corporal because in those days they would never let a coloured fellow be over a white man, not in England, but he was the best educated. He was at university when he joined up.

This, from one of his peers, clearly shows the attitude of the time; on the other hand you can understand the respect the men had for him, not only because he joined up in the Pals but also how educated he was.

OBITUARY FROM YORKSHIRE EVENING POST
FRIDAY June 2nd 1916.
LEEDS " PALS " LOSE AN INDIAN COMRADE.
PRIVATE SEN KILLED IN ACTION
Among the casualties in the Leeds 'Pals' Battalion one is reported today which has a singular interest. A young Indian named J. N. Sen, a native of Chandenagore, Bengal, came to the Leeds University in October, 1910, to study and after taking an engineering course for three years, graduated as bachelor of science. Soon afterwards he acquired a position under the manager of the Leeds Corporation Electric Lighting station in Whitehall, ultimately was placed on the staff as an assistant engineer. While there he gave much promise of a successful career and being of a cheerful disposition, was much liked by everybody. When the 'Pals' Battalion was formed in September 1914, Private Sen, who was then 27 years of age, became one of its first members. He has been killed.

Several months ago when the 'Pals' paraded in the City, Private Sen came in for much notice because of his evident connection with the East. He was a single man and his Mother resides in India. Prior to joining the colours he lodged in Grosvenor Place, Blackman Lane, Leeds, West Yorkshire.

This extraordinary young gentleman, who, with his credentials was capable of considerably more than to be cannon fodder in the appalling front lines of the Western Front, gave his life for the country he knew as home. It therefore brings a tear to the eye when a letter was discovered in 2001, explaining exactly how he died. Until this time he had been forgotten in time and memory. It was a letter written by Private Harold Burniston, a fellow Pal. The excerpt reads as follows.

We suffered a good many casualties ourselves and it was soon after we got back that I heard poor John Sen had been brought in killed. He was hit in the leg and neck by shrapnel and died almost immediately. He was evidently hit in the leg first as when they fetched him in he had a bandage tied round it and must have been bandaging it up when he was hit again in the neck which killed him.

This is part of a much more in-depth letter of the actions of that day and here are just a few excerpts to give an understanding firstly of the levels of education these men had and secondly to understand a little more of the conditions of the trenches in the First World War and Harry Burniston's experiences while on the front lines. Just a note that the man referred to as John was indeed Jogendra Sen; he was known by this name by his friends – perhaps it was easier to pronounce than Jogendra.

Dear Father,
I was very pleased to receive your letter which I got while in the trenches and also one from Mother and Dorothy. What an awful affair the Naval Battle has been; When we first heard of it, it seemed as though we had had by far the worst of it, but we have seen the papers since and are very glad to know that it was not so and that the Germans suffered more heavily than we did, but they were evidently a fairly tough nut to crack all the same.

You say you would like to hear of my experiences so far and about Mr Sen. Well I did not mention it at all in my previous letters as I thought it best not to, however I will tell you what happened on the night Mr Sen got killed.

Our platoon was acting as supports to the first line and were in the trenches just behind the first line trench. We had not to do sentry duty of course, that being done by the men in the first line, but we had to go out every night over the top of the front line trench into "No mans land" putting barbed wire entanglements up in front of our trenches, as you will imagine, it is very risky work and we have to wait until it is getting dark before we go out on top, then get the work done as quick as ever we can and as quietly as we can. As long as it remains dark it is practically impossible for you to be seen from the enemy's trenches, but the trouble is they keep sending blue lights up which burst in the air like rockets and light up everything round about. When you see one of these going up the best thing to do is get down flat before it bursts, but if it bursts before you have time to get down, you stand perfectly still so that they cannot see anything moving about and in all probability you will not be detected as you will simply look like a black figure stuck up and be taken for one of the large stakes stuck in the ground supporting the barbed wire.

It was Monday night the 22nd of May that John was killed and practically all our platoon was out on top wiring. We had been out about an hour and everything had been quiet when all of a sudden we saw several flashes from the other side and then Bang! Bang! Bang! all around us and before we had time to get back into the trench or anything, we were in the thick of it. Well of course the instant it started we were all down flat on the ground and at first I thought that the Germans had spotted us and were sending a few rifle grenades or something over at us, but we soon realised that it wasn't that, but a proper bombardment of our trenches and for about half an hour we were laid absolutely flat. As flat as ever we could get with our faces buried in the grass and our steel helmets on our heads while it absolutely rained shells and explosives of every description all around us. By jove, it was awful and I was expecting to get hit every second. I never expected coming out untouched. Of course they were shelling the whole of our sector of trenches from the front to the rear and I suppose it was practically impossible to walk about the trenches while it was going on, the air being absolutely full of flying shrapnel and everybody got into the dug outs, but it was those shells that were aimed at the front trenches that we were in danger of as we were only about twenty or thirty yards in front of it and every time one burst it went off with a terrific crash and the ground seemed to fairly rock under us and we could hear the bits of shell and earth tinkling on the barbed wire as they hit it. By jove talk about an inferno, I kept having a squint up and everything round was lit up with flames, and the shells were coming screaming over our heads one after another then going off with a crash behind.

Well, about 5 minutes or so after they started, our own artillery started replying back which helped to liven things up even a little more if that was possible and to make matters worse for us laid out on top, began to drop their shells right in front of us, perhaps fifty or a hundred yards or so in between our trenches and the Germans so that the Germans could not get out and rush to attack our trenches without having to come through our artillery screen of fire so we were in between two fires.

Well, as I said, we were laid there about half an hour or so and as things did not appear to be getting any better the corporal in charge of us passed the word down that we had to get back to the trench as quickly as we could so we all commenced to work our way back on our stomachs like eels and it was a job needless to say. We had about thirty yards to do and then we came to a gap in our barbed wire where we had to get through. As we were on the side of it farthest away from our trench of course at this place the wire had a way cut through it so that we could walk through in ordinary circumstances, but still there was any amount of it lying on the ground in the grass and we had to crawl over this for a distance of about fifteen or twenty feet and it was a business as it kept catching in our clothing and we had to pull ourselves free again without getting up from the ground and exposing ourselves more than we could help and in addition we had our rifles with us and a bandolier of cartridges and a bag with a gas helmet in slung over our shoulders. There wasn't half some torn trousers etc when we got back to our trench. Both my trousers knees were torn and the bandolier as well. Talk about heaving a sigh of relief, I did when we tumbled over the top of the parapet into the trench again and the other boys who were inside the trench manning it weren't half glad to see us back safe as we each came tumbling into them. They were all very excited and vowing what they would do to the beggers if they came over to attack us and we all felt the same once we were safely back. I know I did, I felt ready for sticking any Bosch who showed his head over our trench. It was the next day we felt it most after the excitement had gone off.

Well, soon after we got back the bombardment started to slack off and ceased after a do of about three quarters of an hour and immediately it was over two different parties of Germans rushed across with the intentions of making a bombing raid on our trenches but our chaps spotted them in time and opened fire on them with their rifles and drove them off. I believe they suffered pretty heavily too as they left several dead behind them just in front of our parapet and we got four of their dead brought in and captured one of them alive. Our platoon did not take part in driving the beggars back when they came and as soon as we were all safely back in the trench we went back to our own place in the support, but of we had known they were coming I think we should have wanted to stop in the front line and been in at it , I know I should. As of course you will have seen in the papers, we suffered a good many casualties ourselves and it was soon after we got back that I heard poor John Sen had been brought in killed. He was hit in the leg and neck by shrapnel and died almost immediately. He was evidently hit in the leg first. As when they fetched him in he had a bandage tied round it and must have been bandaging it up when he was hit again in the neck which killed him.

He is buried alongside of the other boys who got killed in one of the places where they bury soldiers, just behind the trenches and has a small wood cross over his grave with his name and regiment on. They put one over every grave where possible.

I have not told you before, but even when we are not actually in the trenches and are in billets, we go up to them practically every day as "Working parties" and it is a walk of about three to four miles there and the same back of course and we are either carting great loads of wood, barbed wire, sand bags, etc about up and down the trenches, or else digging new ones and repairing old ones. A lot of the work cannot be done in day time and then we have to go at night of course and are on from about six in the evening until twelve o'clock or later in the morning, so you can tell how much rest there is for the troops when they come out of the trenches. Its all bunkum about rest camps, at least we haven't seen or heard of one yet and we have been here well over three months now.

The last time we were in the trenches the weather was terrible, we were in for eight days and it rained practically all the time and all I and two others had to live in was a hole scooped out of the ground with a piece of corrugated iron over it and we covered the floor with sandbags to sit on. We were not in the firing line this time, but in the reserve and we were carrying the rations from the dump in the fields into the trenches every night. The trenches were in an awful state and over the ankles in water in many parts and the fields outside were like quagmires. We were out every night for about four or five hours and on one or two occasions got wet to the skin. It's a licker to me we don't get laid up with it as our clothes have simply to dry on us, boots as well but we seem to be able to stand anything now. The rations are done up in sandbags which are tied together in pairs by the necks and we sling them over our shoulders and carry them that way. They are awfully heavy and we have to carry them about, Oh nearly a mile I should think from the dump to the trenches and then go back for some more and it generally takes at least two journeys and very often three, So it's some job I can assure you. I pity the poor chaps who were out here at first with the original expeditionary force before there were any trenches dug and things had not been organised as they are now.

Jogendra Sen in his Pals uniform.

Kerton Brothers

When first reading about the Pals Battalion, the first thing that is very apparent is the whole notion of men joining up with their work colleagues, friends and possibly neighbours, but brothers is something which does not occur very often. They may not necessarily be twins or even of similar age but there were many brothers that took to Kitchener's call to arms in 1914. How amazing would that be to two young brothers going on this epic adventure together, many of whom may not have been out of Leeds, let alone to another country. The idea sounds exotic and to be able to share it with your own family must have been a real buzz. It is with great sadness then that this story is retold with the sentiment and respect it deserves.

When the war broke out, Sydney Kerton, who at the time was a master tailor, and his brother Herbert were living with their parents George Henry and Jane Ann Kerton at 26, Hall Grove, Hyde Park, Leeds. Sydney was running his own tailoring business from premises on Woodhouse Lane in Leeds when the Pals Battalion was raised. He quickly wound his business down and joined the Pals battalion on 4 September 1914. His brother Herbert also joined the Pals and was a great addition with his previous army experience serving with the Leeds Rifles, which was a territorial battalion, prior to the outbreak of war. Due to his prior credentials he was instantly made sergeant and soon after Company Quartermaster Sergeant with D Company.

In the early evening of 22 May 1916, while the Pals were on their second spell of trench duty in France, a wiring party with a covering detail were working out in No-Man's Land when they were discovered by a German raiding party. A fight ensued as the Germans tried to enter the Pals' trenches. They were only driven off after Lieutenant Valentine Oland MC had gathered together a bombing party, rallied the wiring party and counter-attacked, driving the Germans into the line of fire of a Lewis gun. (Oland's Military Cross was awarded for this action) Later, when volunteers were searching out in No-Man's Land for wounded members of the wiring party, the German artillery opened up with shrapnel shells, killing and wounding many more men. This minor action cost the Pals fifteen killed, which included the gentle Hindu man Jogendra Sen, and thirty-four wounded including Herbert Kerton and another three missing. Among the dead was Sergeant Sydney Kerton. When daylight came his body could be seen hanging on the barbed wire and his brother Herbert had to be physically restrained from going out to him until later the following evening. It is this that brings the reality of family and war together and how family life for many in the First World War was torn apart; it is however greatly upsetting that Herbert spent the night with his dead brother laying in No-Man's Land and there was nothing he could do. Can you imagine a loved one such as a sibling, parent or even best friend in this situation? This short extract shows just how family life was destroyed when war hit home.

This has been taken from the *Yorkshire Evening Post* of 25 May 1916, shortly after the death of Sydney.

> During the last few weeks Sgt Kerton's wife has been busily preparing a home for her husband, who was expected home on a short furlough. She had taken a house at 6, Glossop Mount, Delph Lane, Woodhouse and had got it practically finished when she received news that her husband had fallen in action.

Herbert Kerton was discharged from active duty in 1919. He had to live out the rest of his life without counselling or help; he just came back from war and got on with his life.

Horace Iles

It is 1914 and you're a 14-year-old lad; just like any other young lad in the early twentieth century, you like to play out with your friends or 'laking out' as colloquially known in the north. Perhaps you have a job, as many young men would be working at around this age in local industry or following in their father's footsteps. To a lot of these young men the prospect of war would never have even crossed their mind; it was older brothers, fathers, uncles and friends who joined the fight for freedom, leaving these young boys at home. One could never imagine a boy as young as fourteen, still at school, joining up for war on his own merit in this day and age, yet in 1914 there were extraordinary cases of this happening; this is the story of one of those boys, who would through his own actions soon become a man.

This is the story of Private Horace Iles. Horace's Father, William Iles, had served under Lord Roberts (Bobs) in the Second Afghan War and William had talked often of his soldiering and Victorian army life. This may have given young Horace the incentive to take the King's Shilling. Leaving School at 13, he worked for a short while as a blacksmith's striker at Pawson's, Marks Road, Woodhouse. Strikers were generally apprentices who had the strenuous task of swinging the blacksmith's hammer to shape the hot scoulding iron. This work clearly filled out his tall frame and by the age of 14 he looked much older than is age. He moved from Pawson's to Dennis & Co., North Lane, Leeds, to train as a painter and decorator soon after the war started and the Pals arrived

The young Horace Iles filling out his uniform. It's not surprising this young man was able to volunteer, much to his sister's dismay.

on Woodhouse Moor with their recruiting tram. Horace added a few years on to his age and was accepted as a new recruit in the prestigious Leeds Pals battalion and sent to Colsterdale for training and, in early 1916, to France. Horace was wounded along with many others in a German barrage of the trenches on 22 May; he was hospitalised in France then given seven days' home leave before he was then sent back to France to join his friends and comrades in the trenches to get ready for the 'Big Push'.

While training at Colsterdale and on the frontlines in France he would as frequently as possible write to his dear sister Florrie; this is one such emotional letter that has not been changed or doctored in any way.

My Dear Horace,
Just a line or two to thank you very much for the card which mother gave me yesterday it is very pretty. I am so glad you are alright so far but I need not tell you what an anxious time I am having on your account, You have dropped in for the thick of it and no mistake, I only hope you have boy I don't care how soon, I should be more than pleased to see you I can tell you.

You have no need to feel ashamed that you joined the Pals now, for by all accounts they have rendered a good account of themselves. No one can call them "Feather bed soldiers" now. I think Barrons boss as got him off so he will feel a bit easier if it is so. He has not told me himself so I am not sure, but I think it must be right. Bob has not heard yet but he is expecting to hear any day now. We did hear that they were fetching all back from France under 19.

For goodness sake Horace tell them how old you are, I am sure they will send you back if they know you are only 16. You have seen quite enough now just chuck it up and try to get back.

You won't fare no worse for it. If you don't do it now you will come back in bits and

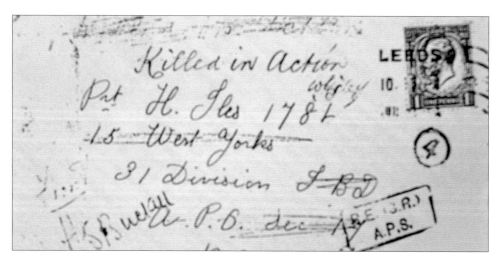

The letter which was returned to Florrie unopened with the moving words, 'Killed in Action,' written across the letter. This was the notification the family was to get that their young Horace had given his life for his country.

we want the whole of you. I don't suppose you can do any letter writing now but just remember I am always thinking of you and hoping for your safe return, So no more this time only my love, Bob says hurry up and come back.

Your loving sister Florrie

XXXXXXXXXXXXXXXXXXXX

Ps. Did you know Roy Mason had been wounded, Shot in the leg, He is now in Manchester Hospital.

This was the last letter that Florrie would send and sadly he was unable to collect and read this letter containing an emotional plea for him to tell them how old he was and to come home. The letter which Florrie sent to Horace and did not receive would have most likely been a reply to what is thought the last letter that Horace would write and send home, the letter is as follows:

Dear Florrie, just a few lines hoping to reach you in the best of health as it leaves me at present. I was discharged from hospital about two days ago, I am sorry I have not written to you before but I had kept putting it off but at last I have written. It is three months since I had a latter from anyone we are serist. Well dear Florrie how are you getting on hope you and the nipper are in the pink. I had a letter from Alice a bit back and she sent her love to you also wanted you to know she is still at dressmaking. I think I shall have to draw to close hoping to have a line soon.
Your loving brother Horace
PS excuse writing pen's bad.

It was on 1 July 1916, shortly after the whistles had been blown and the men climbed the ladders out into No-Man's Land, that he sadly died among hundreds of his friends and comrades. The above letter was not sent until 9 July; as we now know, he had already given his life for his country and his family did not learn of this until the letter was returned unopened, marked 'KILLED IN ACTION'. His family sent a young, enthusiastic, energetic young man to war in 1914 and all they received was this letter to mark his death. It was a fitting tribute to the young man of Woodhouse in the Parish magazine of 1916 which featured this moving obituary.

Horace Iles

Pte Horace Iles of the Leeds Pals was killed on July 1st. He had been a regular scholar and choir boy of our Sunday School. He joined the Army over a year ago though he was but 16 years of age in January last but his build and strength would easily pass him for a lad of 18. Eager to join, he refused to be dissuaded and his letters showed how congenial his preparation was to him. He was a soldier's son and evidently was a soldier born.

Horace with Florrie, his older sister.

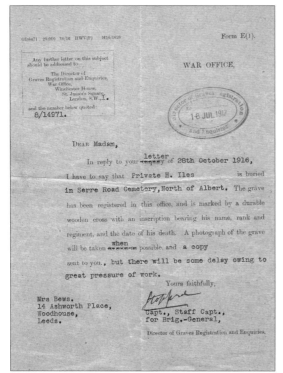

A copy of the letter sent to his family in 1917 denoting where his grave was situated so that his family knew he had been given a proper resting place.

Horace Iles' grave in Serre Road, No.1 Cemetery; in the background are many other graves of brave men that fought alongside Horace in the battle of the Somme.

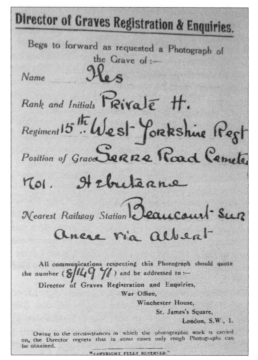

Director of Graves Registration & Enquiries.

Begs to forward as requested a Photograph of the Grave of :—

Name *Iles*

Rank and Initials *Private H.*

Regiment *15th West Yorkshire Regt*

Position of Grave *Serre Road Cemetery*

Rol. Hebuterne

Nearest Railway Station *Beaucourt-sur Anere via albert*

All communications respecting this Photograph should quote the number (8/149 71) and be addressed to :—
Director of Graves Registration and Enquiries,
War Office,
Winchester House,
St. James's Square,
London, S.W, 1.

Owing to the circumstances in which the photographic work is carried on, the Director regrets that in some cases only rough Photographs can be obtained.

"COPYRIGHT FULLY RESERVED."

The letter that would accompany the above photograph showing where the grave is situated in France; his family would receive this upon request.

Later Years and Remembrance

The Leeds Pals Association was set up in 1919 after Colonel Sir Edward Brotherton gave a dinner at the Leeds Town Hall for the surviving members; it was run by a committee elected from its members irrespective of their former ranks.

This would not be their last reunion. The exact details of the founding of the Association can no longer be found, however it is thought that the criteria was that you had to be an original Colsterdale man. It was thought that it was kept this way so that this close knit family would still be able to meet without outsiders joining in. Irrespective of the precise details, we now know that the Leeds Pals Association was founded on the bonds of friendship forged at Colsterdale, in Egypt and on the battlefields of France. A key role of the Association was to take care of the less fortunate members who were disabled as a result of the war. The last president of the association was Arthur Dalby, who served for two and a half years and had been a member of the committee since the beginning.

> I never met a finer body of fellows in my life, and ive always been proud to have been on the Committee and Chairman and President of the Leeds Pals, because they were such a grand lot of fellows.
>
> Arthur Dalby.

> The comradeship continued long after the war, and Fred Naylor always attended the annual outing to Colsterdale, and remembrance service at Leeds Parish Church where the Pal's Memorial is. The real meaning of that comradeship was brought home to me when Grandpa died in 1974 at the grand old age of eighty two. At his funeral I was surprised to see a contingent of Pals stood to their attention outside the crematorium. Then they did a smart right turn in to pay their last respects. This really was a 'Pals' Battalion in the true sense of the word.
>
> Margaret Sudol – Granddaughter of Fred Naylor.

Remembrance is a personal affair and something which differs from person to person; whichever way remembrance takes place, it is the men of the Leeds Pals and many other regiments throughout the country and the men that made these battalions whom we are remembering.

On 1 July 1932, 16 years after the war, surviving members of the Pals stand proudly with their wreath of remembrance and their full chests of medals.

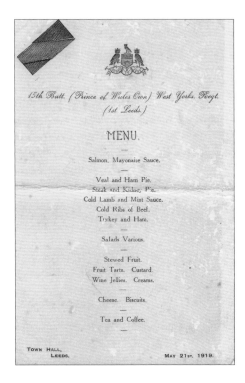

15th Batt. (Prince of Wales Own) West Yorks. Regt.
(1st Leeds.)

MENU.

Salmon. Mayonaise Sauce.

Veal and Ham Pie.
Steak and Kidney Pie.
Cold Lamb and Mint Sauce.
Cold Ribs of Beef.
Turkey and Ham.

Salads Various.

Stewed Fruit.
Fruit Tarts. Custard.
Wine Jellies. Creams.

Cheese. Biscuits.

Tea and Coffee.

TOWN HALL,
LEEDS.   MAY 21ST. 1919.

A menu card from the Leeds Pals Association dinner in 1919, showing the food.

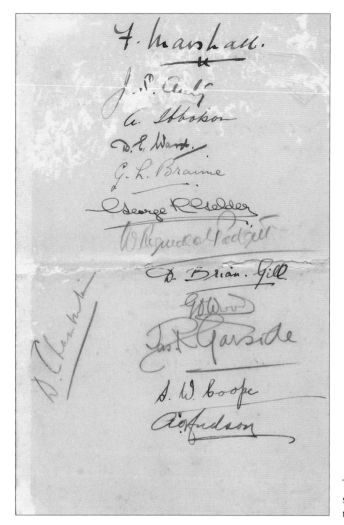

The back of the menu card
signed by some of the Pals at
the meal.

Herbert Bradbourne, one of the last surviving Pals, never spoke of the atrocities
of the war and especially those of the Somme, but every year on Remembrance Day
he would take his wife into his garden and hold his own two minutes' silence to
remember the brave men he fought alongside and some whom he watched lose their
lives. In an interview with the *Yorkshire Evening Post* in an article called 'Somme Hell
Remembered', he was asked what was it like and he gave this reply:

> The food was bad, everything was bad, why relive a terribly memory? We would
> go over the top as though we were going for a walk only to be hit by a battery of
> machine guns, which blew us to hell. One day [1 July 1916] we went over and by
> 9am the Regiment was blown to pieces, 140 survived, the rest were slaughtered, they
> would send us out to face machine guns.

On 2 February 1990, Herbert Bradbourne died a First World War veteran, one of the

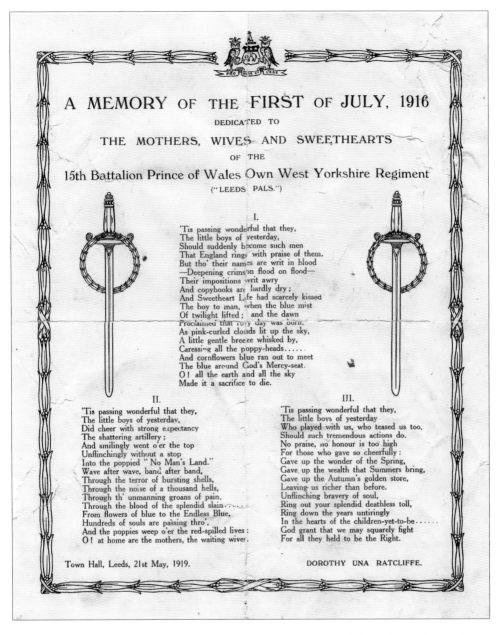

A MEMORY OF THE FIRST OF JULY, 1916

DEDICATED TO

THE MOTHERS, WIVES AND SWEETHEARTS

OF THE

15th Battalion Prince of Wales Own West Yorkshire Regiment

("LEEDS PALS.")

I.

'Tis passing wonderful that they,
The little boys of yesterday,
Should suddenly become such men
That England rings with praise of them.
But tho' their names are writ in blood
—Deepening crimson flood on flood—
Their impositions writ awry
And copybooks are hardly dry;
And Sweetheart Life had scarcely kissed
The boy to man, when the blue mist
Of twilight lifted; and the dawn
Proclaimed that rosy day was born.
As pink-curled clouds lit up the sky,
A little gentle breeze whisked by,
Caressing all the poppy-heads......
And cornflowers blue ran out to meet
The blue around God's Mercy-seat.
O! all the earth and all the sky
Made it a sacrifice to die.

II.

'Tis passing wonderful that they,
The little boys of yesterday,
Did cheer with strong expectancy
The shattering artillery;
And smilingly went o'er the top
Unflinchingly without a stop
Into the poppied " No Man's Land."
Wave after wave, band after band,
Through the terror of bursting shells,
Through the noise of a thousand hells,
Through th' unmanning groans of pain,
Through the blood of the splendid slain......
From flowers of blue to the Endless Blue,
Hundreds of souls are passing thro',
And the poppies weep o'er the red-spilled lives:
O! at home are the mothers, the waiting wives.

III.

'Tis passing wonderful that they,
The little boys of yesterday
Who played with us, who teased us too,
Should such tremendous actions do.
No praise, no honour is too high
For those who gave so cheerfully:
Gave up the wonder of the Spring,
Gave up the wealth that Summers bring,
Gave up the Autumn's golden store,
Leaving us richer than before.
Unflinching bravery of soul,
Ring out your splendid deathless toll,
Ring down the years untiringly
In the hearts of the children-yet-to-be......
God grant that we may squarely fight
For all they held to be the Right.

Town Hall, Leeds, 21st May, 1919.

DOROTHY UNA RATCLIFFE.

A touching memorial poem written and read by Dorothy Una Ratcliffe at the Leeds Pals dinner on 21 May 1919.

last few surviving Leeds Pals and one of the last heroes of the Leeds Pals.

In 1928 the surviving Pals took a pilgrimage to the battlefields of France. For them, it must have been a harrowing yet calming experience, walking around the area where they once fought, lived and saw their friends and comrades gave their lives. Their emotions cannot be portrayed when they arrived in France; all we do know is that it was a unique remembrance for many of them.

The dinner organised by Edward Brotherton was not the last; after this, the remaining Pals took many trips to the battlefields in France where they served and also three annual pilgrimages each year. In 1935 a memorial cairn was placed at Colsterdale, on the site where the original Leeds Pals were sent to train in 1914, where each year a handful of locals go to pay their respects to these great men. On the anniversary of the Pals leaving Leeds for Colsterdale, the Pals Association returned each year to place a wreath as well as placing a wreath on the cenotaph in Leeds.

The front cover of a souvenir guide from a pilgrimage to France by veterans of the Leeds Pals in spring 1928.

Lt-Col Stead at the touching unveiling of the memorial monument at Colsterdale, surrounded by other Pals.

Posing by the bus that would pick them up each year to transport them to Colsterdale on one of their pilgrimages to the area.

Arthur Dalby and Clifford Hollingworth; both are wearing their medals, looking smart, during an interview in 1989.

The Pals at Serre cemetery in 1928, remembering the men around them that they fought with and grew to know.

The Pals reliving memories and walking through some of the preserved trenches at Vimy Ridge in slightly better conditions than those of the war.

The Pals gather for a photo at Serre cemetery.

Surviving members of the Leeds Pals marching up the Headrow in Leeds, past the Town Hall. Perhaps they were singing their marching song: 'We are the Leeds West Yorks/We are the lads who like our beer/We spend our tanners, we know our manners/We are respected where ever we go/And when we march down Briggate/Doors and windows open wide/We can laugh, we can sing/We can do the Highland Fling/We are the Leeds West Yorks.'

These two dates were held so that the memories of their friends and comrades would not be forgotten. The service at Colsterdale is still held today; unfortunately there are no remaining survivors left but family members and others go to keep the memory alive.

It is with great thanks to my father Mike Wood, who campaigned for many years for a memorial to be placed in the heart of Leeds, that there is now a monument of memorial situated outside the Art Gallery to the men of the 15th West Yorkshire Regiment, the Leeds Pals, and is a fitting tribute to the men who fought, served and gave the greatest sacrifice a man can give.

Survivors stood in Ripon cathedral near Colsterdale, on one of their many trips to the area they trained in.

The memorial to the Leeds Pals and Leeds Rifles, a fitting tribute to the men from Leeds, in the grounds outside Leeds Art Gallery.

The Pals rekindling their friendship and sharing memories at the memorial cairn to their lost comrades in Colsterdale.

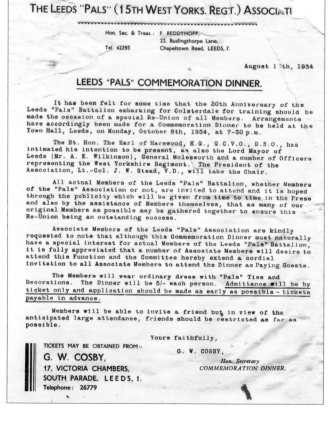

THE LEEDS "PALS" (15TH WEST YORKS. REGT.) ASSOCIATI

Hon. Sec. & Treas.: F. REDDYHOFF,
23, Buslingthorpe Lane,
Tel. 42293   Chapeltown Road, LEEDS, 7.

August 1 th, 1934

LEEDS "PALS" COMMEMORATION DINNER.

It has been felt for some time that the 20th Anniversary of the Leeds "Pals" Battalion embarking for Colsterdale for training should be made the occasion of a special Re-Union of all Members. Arrangements have accordingly been made for a Commemoration Dinner to be held at the Town Hall, Leeds, on Monday, October 8th, 1934, at 7-30 p.m.

The Rt. Hon. The Earl of Harewood, K.G., G.C.V.O., D.S.O., has intimated his intention to be present, as also the Lord Mayor of Leeds (Mr. A. E. Wilkinson), General Molesworth and a number of Officers representing the West Yorkshire Regiment. The President of the Association, Lt.-Col. J. W. Stead, V.D., will take the Chair.

All actual Members of the Leeds "Pals" Battalion, whether Members of the "Pals" Association or not, are invited to attend and it is hoped through the publicity which will be given from time to time in the Press and also by the assistance of Members themselves, that as many of our original Members as possible may be gathered together to ensure this Re-Union being an outstanding success.

Associate Members of the Leeds "Pals" Association are kindly requested to note that although this Commemoration Dinner must naturally have a special interest for actual Members of the Leeds "Pals" Battalion, it is fully appreciated that a number of Associate Members will desire to attend this Function and the Committee hereby extend a cordial invitation to all Associate Members to attend the Dinner as Paying Guests.

The Members will wear ordinary dress with "Pals" Ties and Decorations. The Dinner will be 5/- each person. Admittance will be by ticket only and application should be made as early as possible - tickets payable in advance.

Members will be able to invite a friend but in view of the anticipated large attendance, friends should be restricted as far as possible.

Yours faithfully,

G. W. COSBY,

Hon. Secretary
COMMEMORATION DINNER.

TICKETS MAY BE OBTAINED FROM:-

G. W. COSBY,
17, VICTORIA CHAMBERS,
SOUTH PARADE, LEEDS, 1.
Telephone: 26779

A letter of invitation to a special memorial dinner for the members of the Leeds Pals Association.

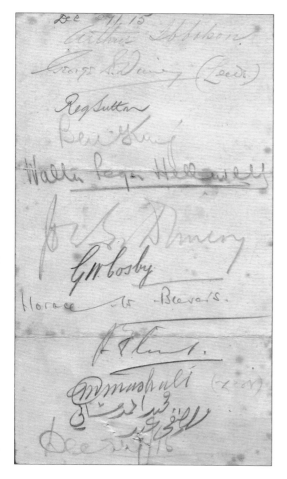

Above left: Another menu card signed; this belonged to Cosby and his signature is recognisable in the centre.

Above right: The cenotaph after its unveiling in November 1922 in the centre of Leeds.

They shall grow not old, as we that are left grow old:
Age shall not weary them, nor the years condemn.
At the going down of the sun and in the morning,
We will remember them

Lawrence Binyon, 'For the Fallen'

UNKNOWN WARRIOR CROSSING CHANNEL ON H·M·S "VERDUN".

The Unknown Warrior crossing the Channel aboard the destroyer HMS *Verdun* on 10 November 1920.